"A FINE SUBSTANTIAL PIECE OF MASONRY"
SCRANTON'S HISTORIC FURNACES

By Daniel K. Perry

Commonwealth of Pennsylvania
Pennsylvania Historical and Museum Commission and
Anthracite Heritage Museum and Iron Furnaces Associates
1994

PHOTO CREDITS
pg. 1 Pennsylvania Historical and Museum Commission; pg. 4 from *Coal, Iron, and Oil*, Daddow and Bannan, 1866; pg. 22 from *Steel and Iron*, Greenwood, 1900; all other images from the collections of the Lackawanna Historical Society, Scranton, PA

ISBN 0-89271-060-8

In Memory of

William Dawson

&

John R. Hennemuth

Acknowledgments

I wish to extend my sincere appreciation to all those who aided in the preparation of this booklet. I want to acknowledge particularly the assistance provided by the staff and volunteers of the Pennsylvania Anthracite Heritage Museum and the members of the Iron Furnaces Committee of the Anthracite Heritage Museum and Iron Furnaces Associates, especially Jack Scheuer and Dominic Keating, and former Anthracite Complex Director Mary Ann Landis. Special thanks to the Lackawanna Historical Society, and Executive Director Maryellen Calemmo for permission to reproduce the images contained herein. Finally, I want to thank my wife Elaine for her continuing help, support and candor.

This booklet has been produced on the occasion of the 150th anniversary of the construction of the first blast furnace in Slocum Hollow in 1840-41.

The Scranton Iron Furnaces Historic Site is administered by the Pennsylvania Historical and Museum Commission in conjunction with the Anthracite Heritage Museum and Iron Furnaces Associates, Inc.

Foreword

The history of Scranton, Pennsylvania, typifies that of an eastern city in America's Industrial Age. The formative period, with its saga of entrepreneurs and immigrants, investors and craftsmen, has attracted the interest of scholars around the country. The sociologist, the geologist, the architect and the archaeologist have been drawn here for the past decade in pursuit of an understanding of the roots of our nation's industrial might.

The general public has joined in this fascination with Scranton. Over 200,000 people journeyed here in 1993 to tour its museums, railroads and mines, and many others came to view its architecture and join in its ethnic festivals. Clearly Scranton's past, purchased dearly with vigor and sweat, is a large part of its future.

What is the genesis of all this? What makes Scranton so typical and yet so different? It is the Iron Furnaces, standing massively to remind the city of its origin and its past as do the pyramids for Egypt or the Mayan temples for Mexico. These large stone, brick and mortar structures gave the United States a product which, perhaps more than any other, served to make our country not a confederation, but a union.

Join historian Dan Perry in this most comprehensive overview at this milestone in our national history. The 150th Anniversary of the Iron Furnaces is an important observance to be sure. Yet more significant is this description of what actually took place. The furnaces shut down in 1902. They were a true product of the nineteenth century. Dan Perry has vividly presented the craftsmanship and resourcefulness of a people of another time. The story of the people who put Scranton on the map and how they accomplished the near impossible are waiting your turn of this page.

Jack Scheuer and Dominic Keating
Scranton, PA
December 1993

Key Dates

1838 — William Henry explores area along Roaring Brook (Slocum Hollow) with the intention of building an iron furnace.

1840 — William Henry, Selden Scranton, George Scranton, and Sanford Grant purchase 503 acres known as the Parson's Lot. Scrantons, Grant, & Co. formed.

1841 — Blast furnace #1 constructed.

1842 — First successful period of blast begins at the "Lackawanna Works."

1844 — Rolling and puddling mill put into operation. Nail making begins.

1846 — Company reorganizes as Scrantons & Platt.

1847 — Rolling of "T" rail begins. Contract signed with New York & Erie RR for a total of 16,000 tons of 58 lb. rail. First stationary steam engine in lower Lackawanna Valley installed.

1848 — Blast furnaces #2 and #3 constructed.

1853 — Firm reorganizes into Lackawanna Iron & Coal Co. and switches from water to steam power to blow furnaces. Blast furnace #4 added.

1857 — Blast furnace #5 constructed.

1872 — Coke-fueled furnace constructed on the site of the original stack.

1875 — Bessemer converters installed. Company begins rolling steel rail.

1881 — William W. Scranton resigns as general manager and founds Scranton Steel Co.

1884 — Lackawanna Iron & Coal renamed Lackawanna Iron & Steel.

1891 — Lackawanna Iron & Steel and Scranton Steel merge and adopt the former's name. Scranton Steel becomes known as the "south works."

1902 — Lackawanna Iron & Steel closes at Scranton. Most of the machinery, tools, and buildings are dismantled and moved to new plant being constructed at Lackawanna, New York.

1969 — Iron Furnaces property acquired by the Commonwealth of Pennsylvania.

1991 — Scranton Iron Furnaces listed on National Register of Historic Places.

"Had the Company known they would not have embarked in this business."

—Selden Scranton to William Henry, February, 1841.

Pennsylvania's Early Iron Industry

Pennsylvania has long been recognized as a major producer of iron and steel. By the mid-eighteenth century, her ironmasters had become highly skilled at casting an amazing quantity and variety of products. On the eve of the American Revolution, Pennsylvania ranked as England's leading iron manufacturing colony. This success was based on the fact that Pennsylvania possessed abundant deposits of readily accessible iron ore, metalurgical-grade limestone for flux, and extensive sources of suitable fuel.

Throughout the eighteenth and well into the nineteenth century, charcoal was the fuel of choice for America's ironmasters. It could be prepared in abundance, was inexpensive, and it provided more than sufficient heat value to

Cornwall Iron Furnace, Lebanon County, Pennsylvania, is characteristic of the Commonwealth's charcoal iron industry. Administered by the Pennsylvania Historical and Museum Commission, the site is preserved and open to the public.

reduce iron ore to its molten state. For these reasons Pennsylvania's early furnaces were organized as "plantations." These typically included thousands of acres of virgin forest from which charcoal was prepared. Old-growth hardwoods were felled at a rate of about an acre a day per furnace. It was the job of colliers to "coal" freshly cut timber in large smoldering mounds—a process that took about two weeks. As had been the case in England decades before, this led to a steady deforestation of southeastern and southcentral Pennsylvania. By the nineteenth century Pennsylvania's ironmasters were looking for an alternative fuel. It was then that several began to experiment seriously with anthracite, or hard coal.

Anthracite Iron

As early as 1777, attempts were made to use anthracite in the smelting process in America. Spurred by the urgencies of the Revolutionary War, the Continental Congress considered a plan to float "stone" coal down the Susquehanna River from the Wyoming Valley to a point near Carlisle in large canoes. The intent was to use anthracite in an air furnace to melt bronze for casting cannon. Although unsuccessful, this experiment clearly illustrates that enterprising businessmen, early on, were seeking an economical alternative to charcoal and were willing to go to great lengths to obtain one.

In the decades that followed, numerous individuals, on both sides of the Atlantic, experimented with everything from preheating the air blast to mixing anthracite with wood, charcoal and coke. Attempts at smelting iron ore with anthracite alone during the 1820s and early 1830s were generally unsuccessful. Each experiment, however, added valuable data to the growing pool of information. In 1838, Joseph Baughman, Julius Guiteau and Henry High blew in a small experimental furnace near Mauch Chunk (present-day Jim Thorpe). It remained in blast for about two weeks and produced several tons of pig iron using air preheated to about 200° Fahrenheit. After several equipment modifications, the furnace was again blown in and remained so until early 1839.

CASTING IRON

From the sixteenth through the nineteenth centuries iron was most commonly smelted in a large truncated stack built of stone. The "blast" furnace, as it became known, was an enclosed vessel. The interior resembled a large upside-down funnel resting atop one which was right side up. At the very bottom sat the hearth where molten iron and slag collected. Just above the hearth several air pipes (initially three and later six at the Scranton Furnaces) penetrated the furnace walls. These were known as tuyeres (pronounced twee-airs) and entered the furnace via archways on the sides and rear of the stack. The archway facing front was known as the casting arch, and is where the molten iron was tapped and flowed into pig beds laid out on a sand floor directly in front.

Into the top of the furnace were poured alternating layers of fuel (coal or charcoal), iron ore and limestone. By 1854, Furnace #3, built in 1848, was capable of holding 1,500 pounds of coal, 1,200 pounds of limestone, and 2,600 pounds of ore at a given time. The limestone acted as a flux, which, when molten, combined with the impurities found in the ore to form slag. Molten slag is lighter than iron, and so it floated on the surface of the iron pool and was tapped off first and discarded. Within the furnace, the three ingredients underwent a steady and continuous process as they descended from the top of the stack to the hearth. Just above the hearth was an area called the bosh. This was the widest, and hottest, section of the furnace. It was in the bosh that the melting action was completed and liquid iron and slag filtered down the last several feet to collect in the hearth.

The furnaces at Scranton held approximately ten to twelve tons of iron in the hearths. When a furnace was judged ready to tap, about every ten hours, the air blast was turned off and a clay plug located in the center of a stone dam was breached with a hammer and chisel. This released the molten iron, which ran into a main runner and then smaller side channels known as pigs. Once cool, the iron was broken into pieces and removed for further processing.

Considered a success by the builders, but probably owing to the insubstantial nature of the machinery, the small furnace was never again lit.

Pottsville's Pioneer Furnace, under the auspices of founder Benjamin Perry, is generally recognized as the first American furnace to operate in excess of ninety days fueled exclusively by anthracite. The use of a preheated air blast and Perry's expertise proved to be the keys to unlocking anthracite's value as a furnace fuel. For this triumph Pioneer's owner, William Lyman, was awarded a prize of $5,000 by prominent native son Nicholas Biddle. Builders of other furnaces, like David Thomas of the Lehigh Crane Iron Company at present-day Catasauqua, quickly followed suit. Anthracite-fueled furnaces soon dominated Pennsylvania's iron industry, particularly in the Lehigh Valley, where good-quality ore, limestone and hard coal existed in close proximity.

Pioneer Furnace

Pound for pound, anthracite burns hotter and longer than charcoal, and because of its density, larger furnace stacks could be utilized without running the risk of crushing the fuel under a heavy burden. In addition to the Lehigh Valley stacks, one of this first generation of furnaces was constructed in northeastern Pennsylvania in a sparsely settled region several miles north of Wilkes-Barre, in what was then northeastern Luzerne County. The small stack attracted little outside attention at first, but within two decades following its inception, the Lackawanna Works had grown to become the second largest iron furnace complex in the nation. This apparent success, however, was not without its share of difficulty.

Birth of Scranton's Furnaces

By 1838, self-described "mineralogist" and engineer William Henry was investigating the feasibility of establishing an anthracite-fueled blast furnace along Roaring Brook in the Lackawanna Valley. He was first attracted to the vicinity in 1836 in connection with a speculative project to extend a canal up the Lackawanna River and link it to a railroad which was to run to Port Colden, New Jersey. The project never materialized because one of the key investors, Lord Charles Augustus Murray (Earl of Dunmore), withdrew his financial support. But Henry was intrigued with what he saw along a section of the proposed right-of-way. Well schooled in the process of making iron, he had been the first American, in 1835, to experiment successfully with applying a hot blast to the smelting of iron ore, at the Oxford Furnace in New Jersey. Henry believed that the Roaring Brook site contained abundant deposits of anthracite coal and iron ore. Writing some twenty years later, Henry reflected that at the time, he had received assurances from "several prominent politicians" that an extension of the North Branch Canal would be constructed to the mouth of Roaring Brook, thus providing a ready source of limestone for flux. Nevertheless, he concluded that these elements, in combination with the fast-flowing and relatively stable waters of Roaring Brook, would provide an ideal location for a furnace.

When William Henry made his initial explorations in "Slocum Hollow" in the late 1830s, the area included a sawmill, gristmill, whiskey still, hotel, cooper shop, and several small dwellings. With the backing of Edward Armstrong, a wealthy Hudson Valley financier who was fond of hunting in the area, Henry negotiated the purchase of the 503 acres known as the "Parsons Lot" in February 1840. Armstrong also had been involved in the aborted railroad/canal endeavor and was particularly interested in exploiting the Valley's coal deposits. The property owners, Zeno Albro, William Ricketson, and William Merrifield, had received title to the land from Ebenezer Slocum's widow. Ebenezer and his brother Benjamin had operated a bloomery on the site from 1800 until 1828, producing small quantities of wrought iron, no doubt utilized by local blacksmiths. By the time Henry toured the area the small furnace had fallen into near total disrepair. Even so, it provided him with evidence that ironmaking in the vicinity had once been feasible.

William Henry

At first the property owners, sensing perhaps an opportunity to make an exceptional profit, asked $15,000 for the parcel. To this gambit Henry balked, and put them off. In the meantime, unbeknownst to the sellers, he further explored the immediate vicinity and concluded that it was even more promising than he had first surmised. The delay paid off, for by the time William Henry met the owners again they had dropped their price to the more realistic figure of $8,000. At this, Henry made arrangements to purchase the tract, pending the arrival of $2,500 from his benefactor as a down payment. Unfortunately, Armstrong died from a horse fall on the day before the draft for the $2,500 arrived. The executor of his estate immediately suspended payment on the property.

On August 5, William Henry notified the property owners of Armstrong's death and his own inability to present a down payment. Securing an extension of thirty days he turned to the proprietors of the Oxford Furnace, his son-in-law Selden Scranton and Selden's brother George. This decision was opportune, for by seeking the Scrantons' backing Henry secured both interested investors and technically proficient ironmasters. The Scrantons, in turn, possessed a familial network which together had the knowledge, ability, financial resources, and dedication to establish an ironworks in such an unproven location.

Above is an artist's rendition of Slocum Hollow as it probably appeared in 1840. First published in 1869 in Horace Hollister's History of the Lackawanna Valley, *this view looks from present-day southside Scranton across Roaring Brook toward central city. The furnaces were constructed immediately in front of the low cliffs to the right of center.*

Scrantons, Grant & Company

By September 8, 1840, the transaction was completed and William Henry, George and Selden Scranton, and Sanford Grant were in possession of the Slocum property. Grant was a Belvidere, New Jersey, businessman who soon became the first manager of the company store. Quickly realizing the need for additional funds, the principles found it necessary to add another partner. Philipp H. Mattes, a German immigrant who was manager and cashier of the Bank of Pennsylvania's Easton branch, purchased a one-fourth interest in the firm for $5,000.

Scrantons, Grant, & Company, as the firm named itself, launched into the anthracite iron business on a financial shoestring. Initially their assets included Mattes' $5,000, $5,000 from Grant and a combined total of $10,000 from the Scranton brothers. In addition, George and Selden turned to cousins Joseph and Erastus Scranton of Augusta, Georgia. By 1843, Joseph and Erastus had invested some $13,000. Henry, on the other hand, supplied none of the initial operating capital, but did provide in-kind managerial services to the firm.

George Scranton

Selden Scranton

Joseph Scranton

Taking up residence in nearby Hyde Park during October 1840, Henry supervised the laying of the blast furnace foundation with the help of local resident Simon Ward and master builder William W. Manness. Reminiscing years later, Ward recalled that no one being present to direct his labor, he borrowed a crowbar from a neighbor and began prying out stone near the site where the first stack was later built. This simple act amounted to the first day's work at the property. Manness eventually supervised the construction of nearly all of the structures associated with the ironworks, as well as the first shops of the Delaware, Lackawanna & Western Railroad. Beginning in late 1840 and continuing through early 1841, a blast stack and necessary support structures rose on the site. When completed, the furnace stood thirty-five feet high with an eight-foot-wide bosh. During this time local ore was mined, construction materials such as fire bricks arrived from New Jersey, and houses were built nearby. According to J. C. Platt in his *Reminiscences of Early History*, the workmen boarded at Mr. Samuel Slocum's tavern at a cost of $1.50 per week. This included bathing privileges and meals. Laborers earned about $17 per month and carpenters were paid at a rate of seventy-five cents per day.

Searching for Ore

Prior to the completion of the furnace, some 3,750 acres of ore-bearing land, "thickly studded with pine, hemlock, and oak timber of original growth," were purchased from the Bank of North America for $11,250. The tract was located approximately three miles southeast of the operation on the southern slope of Moosic Mountain. While Henry had been correct in identifying extensive deposits of coal, he had failed to determine the presence of sufficient quantities of ore in the immediate vicinity. In fact, one of the primary reasons that the Slocums ceased operating their bloomery was the distinct lack of quality ore. The hasty purchase of this additional tract of land was most certainly unexpected and consequently strained the young firm's limited financial resources to the utmost. Construction of the furnace progressed, though, and was completed by Simon Ward and William Manness under Henry's direction in the late spring of 1841. Thomas P. Harper finished construction of the waterwheel shortly thereafter. Unfortunately, the hot-blast equipment did not arrive until early autumn, thus delaying completion of the works. Initially, the new furnace preheated the blast by channeling air through "circular" cast-iron pipes which surrounded the hearth. Air power for this early stack was derived from "blowing tubs," which resembled large barrels in appearance and contained wood pistons with leather rings. Working alternately, they produced steady, although somewhat seasonal, air pressure, for the system was subject to freezing in winter and insufficient water in summer. The tubs were employed until about 1853, when more commonly used iron blowing cylinders were installed and the blast heating apparatus was relocated to the top of the stack.

First Attempts

On September 28, 1841, iron founder Samuel Templin built a small wood fire in the hearth at the base of the stack. Thus began the preliminary job of drying out the interior of the furnace. It was an intentionally slow process, which prevented the new masonry from shrinking too quickly and pulling apart from the intense heat of a full blast.

On October 9, the first "campaign" was initiated, with the stack filled halfway with alternating layers of coal, limestone and iron ore, and the air blast applied. For reasons which are not completely clear, the tuyeres (air nozzles) soon became blocked with semimolten ore and slag that failed to melt completely. In desperation, wood, charcoal, and even brimstone (sulfur) were thrown into the furnace, but to no avail. Within several hours operations were suspended. Workers were then directed to remove a portion of the refractory lining and clear the substantial mass of partially reduced ingredients, called "salamander," which had cooled and solidified within the stack.

Furnaces of the period were tapped every ten to twelve hours as the hearth filled with molten iron and slag. The floating slag was drained off first and later disposed of, followed by several tons of "hot metal." The iron flowed into preformed channels and side channels made of moist sand where it was left to cool and solidify.

A second unsuccessful attempt was made on October 25, 1841. This time the falling water level of Roaring Brook may have been partly to blame. Less water flow meant decreased air pressure in the furnace. Again the tuyeres clogged and operations were terminated. By this point, Henry's competence was being questioned by some of the other partners. These events prompted George Scranton to travel to the Lackawanna Works to determine how the furnace could be fired successfully. After numerous modifications, including alterations to the heating stoves, another attempt was made on December 14. In the meantime, a new iron founder named Clarke from Stanhope, New Jersey, had been hired.

While this attempt was more encouraging than the others, it too ended in failure. With such a dismal record to show for all of the effort and expense, it must have been painfully obvious why local residents referred to the new operation as the "Jersey Humbug." Amidst this desperate situation, Selden Scranton traveled to Danville, Pennsylvania, and secured the services of John F. Davis, a Welsh ironworker, who was familiar with casting anthracite iron.

Lighting the Fire

Davis arrived on January 10, 1842, and immediately proceeded to make alterations. Eight days later the furnace was put into blast and remained so until February 26, when it was "blown out" (shut down) due to equipment problems. During this period approximately seventy-five tons of pig iron had been produced. Although it was an inconsequential amount, even by the standards of the day, the owners and workers struggling at Lackawanna must have hailed it as a rousing victory. Throughout the remainder of 1842, Davis made further modifications to the hearth and blowing equipment,

The deliberate placement of the Lackawanna Furnaces took full advantage of a natural cliffside (which simplified charging from the top), nearby Roaring Brook for power, and adjacent veins of anthracite coal.

and supervised the construction of two new heating ovens. Between May 1842 and March 1843, two more periods of blast occurred. In all, some 975 tons of pig iron were cast, even though the blowing equipment was still inadequate.

The First Workers

Initially, labor was recruited wherever it could be found. During the early 1840s the vicinity was still sparsely inhabited, but workers were soon attracted by the promise of jobs at the furnace. Local residents were joined by an influx of Welsh, Irish, English, and German workers. Contract miners were employed to excavate anthracite and ore and were paid ten cents per ton in 1841. During this early period, experienced miners also were "loaned" from the Scranton brothers' New Jersey operation.

By the second half of the nineteenth century, housing for furnace workers had evolved from rude single dwellings to more extensive row houses. Both types are evident in this view of "Shanty Hill." Vestiges of these structures remain to this day, particularly the block of brick row houses which can be seen from the Central Scranton Expressway as one approaches the downtown.

Ore Mine Road

A gravity railroad ran between the furnace and the Stafford Meadow Brook ore mines, some five miles distant on the southern slope of Moosic Mountain. James Seymour, from present-day Jessup, surveyed the route, and H. H. Easton from Syracuse, New York, was the builder. Empty cars were pulled uphill by mule while full ones coasted back down, powered only by gravity. It was a circuitous journey which, in its infancy, averaged only two loads per day. The ironworks also utilized ore from a mine located several thousand feet above the furnace along Roaring Brook. The former ore averaged only about 25% iron while the latter, a carbonate, yielded almost 50%.

Building a Community

As a means of attracting labor, the company built housing on the hill opposite the ironworks. In 1841, a gentleman by the name of Captain Stott, of Carbondale, was hired to lay out the village of Harrison, as the immediate vicinity had recently been named, in honor of President William Henry Harrison, who had died earlier that year. Rude dwellings were hewn out of the surrounding forest, lending a particularly ramshackle appearance to an area which was soon dubbed "shanty" hill. In an interview conducted in 1916 by *The Scranton Republican,* John Hawks related that "I came here in 1840 with my father [who] had been hired to work at the blast furnaces... All comers had squatters' rights to build on the South Side near the furnaces. They built shanties in the woods and each shanty had a stone chimney. When a newcomer arrived, the neighbors would get together and build a house by sundown." Lumber with which to construct these small dwellings, as well as the wooden furnace-support shops and structures, was produced at the firm's water-powered sawmill located on Roaring Brook across from the furnace. By 1846, the company had constructed some sixty "good and substantial dwellings for the workmen."

The partners intended to produce pig iron, which they hoped to ship via mule wagon to canals at Port Barnum (present-day Pittston) and Honesdale, and then on to rolling and puddling mills along the East Coast. Eventually, they planned to move their iron from the furnace to major eastern markets by rail. Despite the furnace "doing first rate," with several successful periods of blast, the young company continued to face difficulties in 1842 and early 1843. On September 7, 1842, George, writing to Selden from Lackawanna, pleaded that "something must be done speedily for money or we shall be obliged to stop operations." Throughout this period, the price of iron declined steadily. This, combined with the rising cost of transportation, further burdened the company's already strained finances. In fact, shippers, like the Delaware & Hudson Canal Company, were refusing to deliver any goods at all to the Lackawanna Works without cash in advance. This prompted George to threaten that "we will never give them even a box of chocolates to bring again for us," and to observe that "they have served other merchants worse off than us." The bright spot in this rather bleak picture was Lackawanna's company store, which was a busy and profitable operation serving "a store full of customers from morning till evening" and averaging $230 per day "at about 90% profit."

Making Nails

With the prospect of almost certain financial ruin staring them in the face, the partners changed their strategy and conceived a plan to make and sell a finished product. They decided to convert a portion of their cast pigs—they were averaging thirty tons per week at the time—into wrought-iron nails. It was estimated that nails would find a better market and be less expensive to ship. After all, the nation was expanding westward at a rapid rate and even in the East, places that were only villages yesterday were fast becoming cities. The construction of a puddling mill and nailworks, however, required a large additional investment.

WROUGHT IRON

On the average, cast iron contains about 4% carbon along with other elements like sulphur, phosphorous, and manganese. It is the carbon content that distinguishes cast iron, which is hard and brittle, from wrought iron, which is malleable and can be easily shaped by squeezing or hammering. Also, because wrought iron lacks carbon it cannot be hardened or tempered and so it does not hold an edge nor is it very resistent to wear. Probably the most ancient of ferrous materials, wrought iron is actually a combination of two ingredients. This metal includes from 97% to 99% pure iron, the remainder being iron silicate—a glass-like slag which actually helps wrought iron resist oxidation (rust).

Selden and George turned once again to their Georgia cousins. On December 15, 1842, Joseph and Erastus loaned an additional $5,000 to the ironworks. Selden traveled to New York City, where he arranged for a $20,000 loan from merchant-financier John Howland of Howland and Co. In May 1843, the nail factory was under construction on the company's property.

Eventually, pig iron arrived at the rolling and puddling mill via narrow-gauge rail, where it was transformed into wrought-iron T rail, plates, spikes and bar stock. After Lackawanna closed, the rolling mill site was purchased and occupied by the Lackawanna & Wyoming Valley (Laurel Line) Railroad's power house and car barn.

Located along Roaring Brook, approximately 500 yards above the furnace, the new facilities included a rolling and puddling mill, 110 by 114 feet, and a building, 50 by 70 feet, designed to house one spike and twenty nail machines. During April 1844 the rolling and puddling mill was put into operation. It housed five reverberatory furnaces and two trains of rolls, driven by a ninety-horsepower water wheel. The nail factory began production during July 1844. It had a forty-horsepower wheel and an estimated capacity of 100 kegs of nails per day.

PUDDLING IRON

During the period when the Scranton Furnaces were in operation, the most common method of converting cast iron into wrought was referred to as "puddling." The goal was to get rid of as much carbon (and other impurities) as possible, thus reducing the metal to almost pure iron. This was accomplished by reheating the pigs in a reverberatory oven, lined with iron oxide, where they were protected from direct contact with a carbon-based fuel like anthracite. As they were brought to the melting point, the pigs were stirred so that the carbon combined with the oxygen present in the oven, forming carbon monoxide gas. This was carried away by natural draft up a smokestack. As the carbon was removed, the melting point of the iron rose and a large pasty ball formed. After approximately three hours of working the iron, the puddler reached into the oven with a pair of long tongs and removed the spongy "bloom." It was then taken to a large mechanical "squeezer" where excess liquid slag was removed in dramatic fiery spurts. Once reheated, the ball or bloom was further consolidated and hammered into rough bars known as muck bars. These were then bundled together and brought to a white-hot temperature and forge-welded under a large trip hammer into "merchant" bar. After reheating the bar, or billet, it was rolled into a specific shape and size. In the case of nails, it was rolled into flat plates which were later cut into narrow strips, or nail rod.

Puddlers were highly skilled independent contractors responsible for converting cast-iron pigs into wrought iron. The physically demanding process took about four hours in all and yielded a white-hot 300-pound ball of metal, which was removed manually from the furnace with a large pair of tongs.

While the new mills were being readied for production, furnace workers continued to cast and stockpile pig iron in preparation for the day it would be needed to make nail plate. In the meantime, the firm reorganized, and in September 1843 became known as Scrantons & Grant. The active partners were George and Selden Scranton and Sanford Grant. Philipp Mattes, Joseph Scranton, Erastus Scranton and John Howland were listed as special partners whose liability was limited to the extent of their individual investments. Together the Scrantons controlled a little more than 50% of the company's stock. During that same year, George and Selden's younger brother Charles became supervisor of operations. George, who had served untiringly in that capacity, returned home to Oxford, New Jersey, to be reunited with his family.

During the summer of 1844, the furnace averaged five to seven tons of pig iron per day. The nail factory produced so many nails that the owners had trouble selling them all, but this was not the only problem with this item. Due to the low quality of the local ore, the nails, which the Scrantons were making in such large quantities, were very brittle. In fact, many broke when hammered. Consequently, the company's nails met with almost universal complaint.

The Switch to "T" Rail

Steps were taken during 1845 to improve the quality of the iron by mixing ore from Columbia County with local ore. This was successful, and prompted Charles Scranton to describe the new batch of nails as "certainly very fair." The company changed the brand name from Lackawanna to Roaring Brook, no doubt to escape the poor reputation of the former. Unfortunately, by that time the market was again in decline "the nail trade being over done." Even so, the Scrantons put seven new nail-making machines on line by June 1845 in the hope that the market would improve. It did not. Faced with mounting debts and no way to meet them, the company's officers decided to augment the manufacture of nails with "T" rail for the railroads. In a letter dated November 5, 1845, George advised Selden that "we must begin to think about making R. R. Iron rail. I am now satisfied that it's going to be a great business for 15 years with great prices."

During the 1830s and 1840s, America's railroads expanded at an amazing pace. Thousands of miles of new track were being laid annually. Virtually all of this rail, however, was imported from England. It was expensive and sometimes difficult to obtain as Britain's ironmasters enjoyed their monopoly. Consequently, a great opportunity to enter the market was offered to American furnaces. Documentary evidence indicates that the first American T rail was rolled at the Mount Savage Ironworks in Maryland in 1845. With the great boom in railroad construction that was occurring, other American ironmasters quickly followed suit.

In an effort to gain familiarity with this new process, Selden and Charles visited the Montour Works in Danville, Pennsylvania, in early 1846. George traveled to New York City to find financial backers for this latest endeavor. Although he found few investors initially, he did come into contact with Benjamin Loder, president of the New York & Erie Railroad.

A Railroad Is Saved

Loder's railroad, under the terms of a relief act passed by the New York legislature in 1845, was required to complete a connection from Piermont on the Hudson River northwest to Binghamton by December 31, 1848, or forfeit a state subvention of $3,000,000. Faced with obtaining an uncertain and expensive supply of English rails, Loder became interested in the planned expansion at Lackawanna. As far as the Scrantons were concerned it was simply a case of being in the right place at the right time.

On September 16, 1846, the Scrantons negotiated a contract with the New York & Erie for the manufacture of 4,000 tons of 58 lb. (weight per yard) T rail at $65 per ton. Furthermore, they engineered a loan with the company to expand their works. The railroad sent Loder and board member William E. Dodge to inspect the ironworks on Roaring Brook in October 1846. They and eight of their associates subsequently advanced $90,000 to Scrantons & Grant as well as agreeing to an additional contract for 12,000 tons of rail.

The Iron Company Reorganizes

In November 1846, the company reorganized as Scrantons & Platt. Joseph C. Platt, the son-in-law of Joseph Scranton, took the place of Sanford Grant, who had tired of the iron business. Platt became manager of the company store and general buyer. He later was appointed real estate agent. The

Joseph C. Platt

new company listed capital stock at $230,000. By this time, the firm also owned over 5,000 acres of land, 1,200 of which were underlain with coal. In fact, it was the presence of coal that most interested many of Lackawanna's most ardent financial backers.

Coupled with the increase in capital and assets the company also realized a substantial growth of prestige. The new investors from the New York & Erie Railroad were some of the wealthiest and most powerful men in the nation. George W., Joseph H., and Selden T. Scranton, and Joseph C. Platt were listed as general partners. Philipp H. Mattes, Edward Mowry, John Howland, William E. Dodge, Anson G. Phelps, Benjamin Loder, Samuel Marsh, Henry Shelden, John I. Blair, James Blair, William B. Skidmore, James Stokes, Philip Dater, Daniel S. Miller, John A. Robinson, William H. Shelden and Frederick Griffing became special partners. By October 2, 1847, the company's capital stood at $250,000. Imbued with a renewed spirit, George Scranton boasted that Lackawanna would soon be the strongest ironworks in America.

The rolling and puddling mill was expanded to meet the new orders for rail. This included a 110-foot addition containing eight more puddling ovens and an eighty-horsepower steam engine "with trains attached sufficient to finish at least twenty five tons of Rail Bar per day." The engine was operational by late July, and on August 9, the company began rolling rail under the auspices of the Rev. John R. Williams, pastor of the First Welsh Calvinistic Church, and his assistant and fellow Welshman Edward Coslett. Williams was described as an expert iron roller. Scrantons & Platt delivered their product at various points along the New York & Erie right-of-way by wagons pulled by oxen and mules. These timely deliveries enabled the railroad company to complete the line to Binghamton with only four days to spare. By striking this singular bargain, both the Erie Railroad and the Lackawanna Ironworks were spared from bankruptcy and placed on the road to financial success. In fact, with the completion of the line as far as Binghamton, the Erie, for a brief period, was the longest railroad in America.

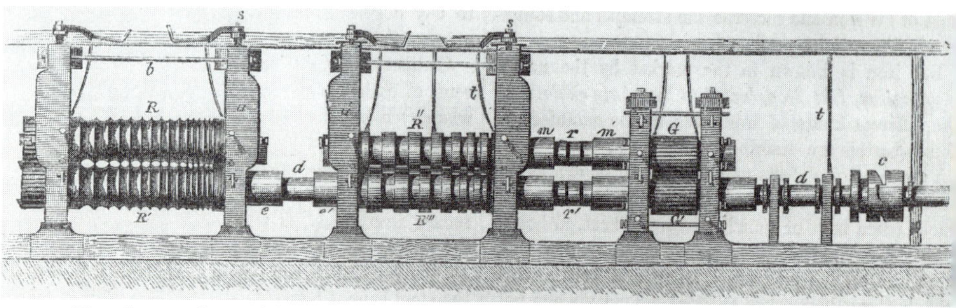

T rail was formed by squeezing billets of hot iron between heavy cast-iron rolls. Multiple passes altered the profile until a desired shape was attained. Lackawanna both purchased as well as produced its own rolls, and other heavy parts, at a company-owned foundry and machine shop.

The Lackawanna ironworks continued to prosper and expand. In late 1847, the company received an award from the American Institute for producing the best specimen of T rail in the nation.

By 1847 the company listed over 800 employees. An amazing growth in just seven years! Welsh, Irish, and German immigrants lived just across Roaring Brook on Shanty Hill. In an interview published in *The Scranton Republican* in 1903, German immigrant John Scheuer recalled walking from New York City to Bucktown (Dunmore) in 1849 where he was introduced to Selden Scranton. "Mr. Scranton made us show him the palms of our hands to prove that we could work... Henry Hess and I got work at the blast furnaces. Men were scarce, and we had to do all kinds of work. My first work was taking out the cinders. Later I hauled props to the ore mines and supplies to the old rolling mill."

In 1850-51 the plan of the city of Scranton was laid out by Joel Amsden, an architect hired by the company. As the success of the ironworks increased, so too did the number of immigrants coming to the area to seek employment. During 1848-49, two more furnaces were constructed; by 1852, more than 3,000 people lived in the immediate vicinity.

The inception of the anthracite iron industry, a landmark in American ironmaking history, reflected a shift from a predominantly rural industry to one that was to become increasingly urbanized as the size of the individual plants and the need for workers expanded dramatically. This was true, not only at Lackawanna, but throughout the Lehigh Valley, and at Danville. Anthracite ironworks served as the corporate model for the massive, integrated steel plants which evolved in places such as Pittsburgh and Bethlehem.

Birth of the D.L.&W. Railroad

During the summer of 1850 the company began work on the Liggett's Gap Railroad. This was a major step toward improving the transportation links into and out of the lower Lackawanna Valley. Under the driving force of George Scranton, the road eventually connected the ironworks to the New York & Erie Railroad at Great Bend, a distance of forty-eight miles. By April 14, 1851, the renamed Delaware & Western Railroad Company was hauling iron ore and limestone to the furnaces and bar iron and T rail to market. Soon after, a second right-of-way was constructed which ran in the opposite direction, east through Cobb's Gap. This line offered rail access to markets in New York City and Philadelphia and provided a connection to the rich magnetite ore deposits of northern New Jersey. These two railroads quickly

THE CITY OF SCRANTON

Scranton's foundation was built of cast iron and its framework of rolled steel. While the establishment of the furnaces predated the widespread mining of coal in the area, the success of the Lackawanna Ironworks created a city destined to become the "Anthracite Capital of the World."

Within the brief span of twenty-five years an incorporated city of 20,000 residents stood in a place that had, prior to the establishment of the furnaces, contained a few isolated hamlets and dwellings. The city's future, although tied closely to ironmaking, was not entirely dependent upon it. From the construction of the first furnace onward, the Scrantons and their business associates welcomed investors and entrepreneurs. More so than most established eastern cities, Scranton was open to newcomers willing to gamble their prospects on a wide variety of opportunities. By 1904, two years after the furnaces closed, Scranton boasted 116,000 inhabitants, five steam railroads, forty public schools, eighty-two churches, thirteen banks, one hundred forty manufacturing interests and thirty coal mines. During the last quarter of the nineteenth century, anthracite mining had supplanted the manufacture of iron and steel rails as Scranton's leading industry. These two endeavors, together with railroading and textile production, made for a strong economy. When the furnaces closed for good in 1902, the city was shaken, but not devastated. The most significant consequence of the closing was that Scranton became increasingly reliant on coal. As anthracite began losing its share of America's energy market during the first half of the twentieth century Scranton's economy faltered. With the collapse of the coal industry the city was forced to diversify in order to survive. Interestingly, it was Worthington Scranton, grandson of Joseph Scranton, who conceived of a bootstrap plan to revitalize business and industry in the city that bears his family's name. Initiated in 1945, The Scranton Plan was, and continues to be, a very successful public/private partnership designed to promote business and industry in the greater Scranton area and to provide low interest loans through SLIBCO (Scranton-Lackawanna Industrial Building Company) for development.

merged to form the Delaware, Lackawanna & Western. With the construction of these roads, the valley's remaining coal lands were quickly purchased in large parcels and the widespread mining of anthracite in the valley began. Due to the foresight of Lackawanna's partners, particularly the New Yorkers, these men were to reap immense profits in the coming years from their coal lands and the hauling of coal.

Large brick ovens, called stoves, sitting on top of the furnaces captured waste gasses, which were used to preheat fresh air pumped into the stacks below. One of the side tuyere arches of furnace #1, abandoned by the time this photograph was taken in the 1860s, is visible behind the casting house.

Lackawanna Iron & Coal Company

In 1853, Scrantons & Platt reorganized again and became known as the Lackawanna Iron & Coal Company. The company's assets included three furnaces, the rolling and puddling mill, a foundry, two blacksmith shops, a car shop, two carpenter's shops, a sawmill, a gristmill, an office, a company store, 200 houses and dwellings, a boarding house, assorted officers' houses, ore and coal mines, a tavern and a recently completed hotel. Some 8,000 shares of stock were issued and the firm's capital rose from $400,000 to $800,000. Selden Scranton was the first president, Theodore Sturges became the treasurer, Joseph H. Scranton was appointed general superintendent, and Moses W. Scott served as both secretary and assistant treasurer. Joseph C. Platt continued to serve in the capacity of storekeeper. Other directors included Samuel Marsh, Daniel S. Miller, Lothrop L. Sturges, John I. Blair, William E. Dodge, John Howland, and George W. Scranton.

The large structure at the center of the photo housed Lackawanna's massive steam-powered blowing engines. Air was pumped from there to the waiting furnaces below. The original company store and office stand in front of the engine house.

Harnessing Steam Power

In 1847, the owners began the conversion from water to steam power. The first engine was installed at the rolling mill and used to power one of two trains of rolls. It was the earliest stationary steam engine to be employed in the lower Lackawanna Valley and was followed by blowing engines at the furnaces in 1853-54. The two double-connected lever-beam engines installed at the furnaces were considered the largest of that type ever used in America. Built by I. P. Morris

& Company, of Philadelphia, the flywheels alone weighed 40,000 pounds and were ten feet in diameter. Within three years, two more engines were installed at Lackawanna. By 1879, a total of seven steam engines were in operation at the furnaces, producing more than 77,000 cubic feet of air per minute at a pressure of between five and seven pounds per square inch. The engine house was located above and behind the top of the furnaces. Air was pumped via a single large pipe, known as a downcomer, to the furnaces below. A manifold directed smaller branch pipes to the individual tuyeres.

Lackawanna employed small steam locomotives, known as "lokies," to transport raw materials, pig iron, heavy machine parts, etc., between the various company operations. These were the company's workhorses and ran on a nearly continuous basis.

In many ways, Lackawanna Iron & Coal took full advantage of American technological innovation during the mid-nineteenth century. Among the first in the nation to smelt iron with anthracite, the company's proprietors were constantly modifying the furnaces and mills in an effort to increase production. In 1854, Lackawanna boasted the third largest rail mill in the nation. By 1865, Lackawanna possessed the capacity to manufacture 60,000 tons of iron annually—more than any other furnace complex in the country. Even though the Scranton Furnaces had such awesome potential, they never produced iron at, or even near, capacity. For example, in 1865 the Lehigh Crane Ironworks of Catasauqua possessed a capacity of 50,000 tons annually, and in actuality produced nearly 42,000 tons during that same year. Likewise the Thomas Ironworks, also of the Lehigh Valley, had the estimated capacity to manufacture 50,000 tons, and missed that amount in 1865 by only about 1,000 tons! Part of the problem at Lackawanna during this period was that furnace #5, the newest and largest stack, was in blast for only twenty weeks.

From its earliest days, the Lackawanna Ironworks maintained an extensive coal mining operation. Throughout the nineteenth century the company used all of the coal that it mined in its ironmaking. Pictured here is one of the earliest coal-processing plants, or breakers, constructed in Scranton. It was built by the iron company and was known as the Capouse Colliery.

A New Type of Furnace

In 1872, the company constructed a blast furnace which was radically different from the others and patterned after furnaces then in use in western Pennsylvania. While the previous furnaces were made of stone and fueled by anthracite, the new one was sheathed with riveted iron plates and used coke for fuel. Coke is a derivative of soft (bituminous) coal and thus had to be shipped in, much to the dismay of local miners who viewed it as a threat to their livelihoods. The company's reasons for building this furnace were several. By this time, coke was well established as an efficient and cost-effective furnace fuel. Although the soft coal had to be transported to Scranton, the new furnace possessed a much greater capacity for producing iron than the earlier stacks. When constructed, it was regarded as the largest furnace in America and by 1879 had achieved an output of 629 tons per week. The use of an alternative fuel made it possible for Lackawanna's ironmasters to continue casting even during prolonged hard coal strikes.

The 1872 coke furnace was located on the site of the original furnace and is shown here under construction. The new furnace was encased in iron plates and possessed a significantly greater capacity than the earlier, much smaller stone stacks.

Taking Stock

By 1880, Lackawanna's assets included five blast furnaces; a rolling mill housing 113 puddling and 35 heating ovens; a steam-powered sawmill; a gristmill; a foundry which averaged 500 tons of castings weekly; machine, car, carpenter, harness and wheelwright shops; a brickyard; and an extensive dry goods store together with company offices. This last structure was completed in 1868 and stood at the corner of Lackawanna and Jefferson avenues. Ore for the stacks came into Scranton from the company's mines at Mt. Hope, New Jersey; Brewster, New York; and Dover and Franklin, New Jersey. During 1880, the Scranton operation produced about 125,000 tons of pig iron while an additional 25,000 tons were cast at the company's furnace at Franklin, New Jersey. This total was subsequently converted into 50,000 tons of iron rail, 80,000 tons of steel rail and 5,000 tons of merchant iron.

A new and greatly enlarged company store was constructed in 1868 at the corner of Lackawanna and Jefferson avenues. It sold a wide variety of dry goods to employees as well as the general public. The structure also housed Lackawanna's corporate offices.

Men and Machines

Like other heavy industry of the period, Lackawanna experienced its share of labor unrest. Concern for the welfare of employees was tempered by a driving desire to produce iron in ever-increasing quantities. Skilled employees, such as founders and puddlers, were generally well compensated and given incentives to stay with the company. This, however, was tied directly to the fortunes of the firm, and on at least one occasion the board of directors attempted to reduce the salaries of the skilled employees and supervisors by one-half in the face of a severe decline in business. Laborers, like guttermen and fillers, endured difficult and dangerous conditions in exchange for low wages and long hours. Interestingly, in 1850 superintendent Joseph Scranton assented to a request by the workmen that on the Sabbath "certain kinds of work which may have been deemed necessary but upon due reflection, we are satisfied is unnecessary" not be performed. Although it is unclear which employees were given Sundays off, it would certainly not have been those who were directly responsible for maintaining the furnace fires and related machinery.

As with most furnaces, advances in ironmaking technology played an important role at Lackawanna. Information on improving production was gathered from many sources. As early as 1848, Selden Scranton was instructing his brother that "if you meet any iron men I want you to compare notes." In 1857, Joseph H. Scranton, still serving in the capacity of superintendent, traveled to the Cambria Ironworks at Johnstown and noted some of the site's more innovative designs. The prominent Welsh ironmaster David Thomas, of the Thomas Iron Works, visited the Scranton furnaces in 1856 and made a number of recommendations including altering the bosh dimensions (suggestions that not only were carried out but that ultimately improved output). William W. Scranton, son of Joseph H. Scranton, traveled to Europe in

The workers captured in this image are guttermen, who were responsible for tapping their furnace (#5), laying out the pig beds, and loading the solidified pigs onto rail cars. This is one of the few known photographs that highlight company employees.

1872 to study English, French, and German iron and steel-making techniques. The most significant technological improvement at Scranton, however, was probably the installation of Bessemer converters.

Making Steel Rail

The first steel made in America using the Bessemer process was made in 1864 at Wyandotte, Michigan. Nevertheless, by 1871 only five American mills used this technique. On June 16, 1874, Lackawanna Iron & Coal laid the foundation for two Bessemer converters—the tenth company in the United States to do so. On December 29, 1875, the firm began rolling steel rails. The new steel mill included a cupola room housing four cupola furnaces, each capable of melting five tons of pig iron in thirty minutes, and two ten-ton ladles mounted on scales which carried the hot metal to the hoisting towers located at each end of the room. The adjacent converter room had a floor that measured 84 by 124 feet and a ceiling 21 feet high, and it contained the two 15-foot-tall converters capable of holding seven tons of steel apiece. Air for the converters was provided by two 500-horsepower, independent, horizontal blowing engines manufactured by the Dickson Manufacturing Company, Scranton. Each engine was capable of delivering 9,500 cubic feet of air per minute at a pressure of twenty pounds per square inch.

The Bessemer steel works and rolling mill #2 were constructed on the low bluff on the side of Cedar Avenue opposite the blast furnaces. The site was later occupied by the locomotive shops of the D.L.&W. Railroad and today houses a factory that produces various types of ammunition for the U. S. Army.

BESSEMER STEEL

The "Bessemer" process, developed independently by Englishman Henry Bessemer and American William Kelly during the 1850s, involved the passing of oxygen under pressure through the bottom of a large refractory-lined egg-shaped vessel filled with several tons of molten pig iron. The resulting "blows" burned off excess carbon in a dramatic shower of flame and sparks. A controlled amount of carbon was then reintroduced to the nearly pure iron in the form of molten "speigeleisen." The contents of the vessel were tapped off from the bottom into waiting ingot molds. Lackawanna's cast-steel ingots varied in size depending on the size of the rail being rolled. For example, sixty-five-pound rail thirty feet long was produced from an ingot which was twelve inches by twelve inches by forty-five inches long. Ingots were reheated in below-the-floor "soaking pits" prior to being passed between huge rolls. With the development of the Bessemer process, steel production, for the first time in history, was measured by the ton rather than the pound.

Scranton Steel Company

In 1881, following a disagreement with Lackawanna's board of directors, William Scranton left the firm and with the help of his brother Walter founded the Scranton Steel Company. Charles F. Mattes, son of original partner Philipp Mattes, was appointed in his place. Scranton Steel was located along the Lackawanna River about one mile below the Lackawanna Works. They rolled their first steel rail on May 4, 1883. By 1890, Scranton Steel's two six-ton Bessemer converters were producing 250,000 net tons of ingots per year. These were rolled into some 220,000 net tons of steel rails. This competition ended in 1891 when the two companies merged to form the titan Lackawanna Iron and Steel Company, the third largest steel works in America. The former Scranton Steel plant became known as the "south works." In 1894, the new company manufactured 500,000 net tons of steel rail. This quantity of T rail was one-sixth the total national output and equal to one-third that produced in England at the time.

Like the Lackawanna steel works, Scranton Steel boasted two six-ton-capacity Bessemer converters, several large cupola furnaces, and an extensive rolling mill. They did not produce their own pig iron, however, but relied on outside suppliers to deliver it by rail.

While the blast furnaces at Scranton produced pig iron at an unprecedented rate, it was still insufficient to satisfy the appetite of the company's steel converters. Although by 1898 only two of the Scranton furnaces were listed as active, the company operated five furnaces in Lebanon County, Pennsylvania, and one at Franklin, New Jersey. These produced Bessemer pig iron, which was shipped to Scranton and converted into steel rail.

Distant Sources of Ore

Ore was also being hauled to Scranton from some distance away. Due to the poor quality of the local ore—David Thomas referred to it as "very lean and hardly worth working"—Lackawanna's ironmasters found it necessary, early on, to locate other sources. By the last quarter of the century, thousands of tons of ore were being shipped to the Furnaces by rail from New Jersey's Mount Hope Mines, the Cornwall Ore Banks of Lebanon County, Pennsylvania, and the Foster Tilley Mines of upstate New York. Blended together, these ores were converted into a very high grade of low-phosphorous pig iron. The transport of vast quantities of ore to Scranton, however, played a significant role in the eventual relocation of the Lackawanna Works.

Beginning of the End

During the second half of the nineteenth century, the center of America's iron industry began to shift west following both raw materials and customers. Men such as Andrew Carnegie built huge integrated mills in the Pittsburgh area to take advantage of navigable waterways, and abundant sources of soft coal for coke. Others, were looking to capitalize on the newly discovered ore deposits of the Upper Midwest. During the 1890s, Carnegie waged an all-out price war for the T rail trade. By underselling his competitors, including Lackawanna, he tried to force them out of business. Because of the high cost of shipping iron ore into Scranton as well as changing markets, Lackawanna Iron & Steel found it difficult to compete and so in 1899 the board of directors decided to move the plant to Buffalo, New York. By constructing a new plant there, they felt that they would be able to take advantage of the ore boats that plied the Great Lakes. Also, by the late nineteenth century, the Scranton operation had become surrounded by urban development which made future expansion unlikely.

During the last quarter of the nineteenth century the New York City faction on the board of directors gained a majority of the firm's stock. Consequently, the decision to move the plant was made by individuals living elsewhere who based their plans solely on business prospects with very little consideration given to the impact that it would have on the city. Despite a valiant effort by the Scranton Board of Trade to convince the owners not to close, the plant, according to

When Lackawanna Iron & Steel Company relocated to Buffalo, it left the imposing stone stacks behind. They have withstood the ravages of time and the elements, and today stand as a monument to the industry and family which founded the city.

one bitter observer, was "deliberately butchered in cold blood." Most of it was shipped to what became Lackawanna, New York, on Lake Erie. The firm's two coal mine shafts, Pine Brook and Briggs, were sold to the Scranton Coal Company. Their other mine, Rolling Mill Slope, had been abandoned long before the decision to move. By 1902, nearly all of the structures which had for so long occupied the banks of Roaring Brook were gone along with more than 2,500 jobs. All that remained were the technologically out-dated stone blast furnaces.

During the sixty years that the Lackawanna Furnaces produced iron, they were the single most significant factor in the economic, social, and industrial life of Scranton. Local historian Frederick L. Hitchcock in his *History of Scranton* estimates that the "invested" value of Lackawanna at the time of its closing was around $6,000,000. When operating at its maximum the Lackawanna Iron & Steel Company employed from 3,000 to 5,000 men at its furnaces, mills, mines and shops.

Founded in a rugged wilderness, and lacking in the immediate vicinity two of the three essential ingredients for making iron, Lackawanna's founders succeeded where many others would have given up. They did so with vision, energy, and a willingness to gamble in the face of ruin. When they were unable on three occasions to light their new furnace, they found someone who could. When the manufacture of nails proved inadequate they expanded the size of the plant dramatically and pegged their fortunes to T rail—delivering it not by canal or locomotive but by mule wagon! This was followed shortly thereafter with the construction of their own railroad which linked their outpost with both ready markets and much needed resources. In founding an iron furnace the Scrantons founded a city.

Lackawanna helped to make Scranton the largest city in northeastern Pennsylvania. For years the city was recognized nationally as a center of iron and steel production, playing a major role in America's industrialization. The company's development propelled the expansion of the Lackawanna Valley's anthracite industry, which in turn prompted growth in railroading and textile manufacture and earned for Scranton a significant place in the history of America's Industrial Revolution.

An Account of the Lackawanna Ironworks

excerpts from *The Wilkes-Barre Advocate*
November 3, 1852

In the year 1840 the Lackawanna ran thro' an almost unbroken forest, disturbed only by the sounds of labor near the openings of a few small coal mines, and the occasional clink of iron around the rude furnaces of the brothers Slocum, after whom the locality was then called Slocum Hollow.

The Slocums were succeeded by Mr. Henry of Wyoming, (father in law to Mr. Selden Scranton,) who erected two furnaces and commenced the settlement. He was succeeded by the Company now owning the entire place, and all the mills and works. Messrs. Scranton & Platt, who form a limited copartnership, the stock being owned extensively in the chief eastern cities...it will be singular if many more years elapse without its producing such an effect on the iron and coal trade as to make it known everywhere. It now numbers over 3,000 inhabitants,—all or nearly all in the employ of the Company, in the furnaces, rolling mill, foundry, car shop, steam saw mill, steam flour mill, iron or coal mines, or in one or another of the busy places with which the village abounds.

The wealth of Scranton lies in the earth underneath it, which is full of iron and coal. The Lackawanna flows through a valley from three to ten miles wide, and the Roaring Brook joins it at the village, affording a small water power for minor purposes, the main power being steam. I counted six steam engines today which I saw working. The convenience of the mines is a great matter of surprise and admiration to the visitor. The boulder ore is brought by railroad some three miles out of the mountain. The coal is everywhere. Fifty openings meet your eye, with rail tracks coming out of them, all within fifty rods of the furnaces, and some of them within fifty feet. One large coal mine opens directly between the stacks of the furnaces, and you pass into it through the buildings.

It is difficult to say which is likely to be the greatest source of wealth, the coal or the iron. The Liggett's Gap Railroad has commenced a revolution in the Western coal markets, and I was very much surprised in coming on from Dunkirk to observe the long trains of coal which were going westward, winding their snake like lengths along the banks of the Susquehanna and Allegheny. The introduction of Anthracite into the Western market is a matter of vastly greater importance than is supposed at first thought. To household use it is of comparatively small moment; but on navigation of the lakes it is producing a great effect. Many of the large steamers have already begun to use Scranton coal, and by another year it is probable that no first class steamer on Lake Erie will burn anything else. The facility with which steam is kept up, enables the boats to surpass in speed all the others, though the same boats have heretofore been the slowest of their class on the lake.

The early completion of the Delaware and Cobb's Gap Railroad will produce a still greater effect on the Eastern markets. Few persons know how near New York is to the great coal fields of Pennsylvania, and few will know it until they are startled by the first five hour train from Scranton to the great metropolis over the new road connecting at Hampton (below Easton) with the New Jersey road. It appears to me that the construction of this railroad is likely to be of greater importance to New York and the eastern cities, than that of any road now projected. For, although it will bring nothing but coal, that single article is now the source of nine-tenths of the wealth of the seaboard. It propels our ships, drives our factories, and is in fact the slave of all our toil, and the producer of all our wishes...

But though a mining country is proverbially poor, this valley will never be so, because the earth in affording coal at the same time furnishes iron, so that the means of consuming coal for a useful and wealth-creating purpose, is at hand. There is no waste of the convenience of nature here. The extent of the works can only be known by a thorough examination of them; and I propose to give a detailed history of a bar of iron from the time it leaves (in the pig) to the last touch of the hand that finishes it. The furnaces stand in the bank of the ravine, through which Roaring Brook once ran, but in which it is now dammed; and we were called by the sound of the bell to see the tapping of one of them, and the liquid iron running into the pigs, where we left it cooling and changing color a hundred times as it cooled, while we walked some rods up the ravine to the rolling mill, first entering the stack of the...

PUDDLING FURNACES

Into these the pigs of iron (broken into pieces) are thrown, and subjected to a strong blast of heat, which in about forty minutes reduces them to a boiling, foaming mass, bubbling up and subsiding as the workmen stir it with long bars of iron. It would seem that such glow and heat must burn the eyes out of their heads. But they seem to have good vision, nevertheless. Looking through a small opening in the door, they constantly stir the foaming iron until after an hour or less, it commences a process precisely like the collecting of butter in a churn. The men continue their care collecting the lumps as a dairy woman does the butter, until they have a large mass of iron, mixed of course with some cinder and dross, and at length this compact lump weighing from four to five hundred pounds, is drawn out of the furnace door, glowing like a small sun, and goes rolling and shining down an inclined plane into the rolling mill, where it is seized and placed in the constantly opening and shutting mouth of...

THE PINCHER

or squeezer. This is a large instrument, looking like a huge crocodile chewing a mass of iron. A skillful workman, rolling this mass over and over, and turning it this way and that way, subjects it to the powerful pressure of the jaw on all sides, while at each pressure small jets of fiery liquid pour out, and run dripping down from the lower jaw to the ground. When the Pincher has pressed out all the dross and cinder which it has power to do, the mass, shaped like a cylinder but rather smaller at one end than the other end, is deftly whirled across the floor, and caught by other workmen, who immediately pass it through...

THE ROLLERS

Of course every reader knows that these are heavy iron cylinders, turning with tremendous force and rapidity, one touching the other, and having a series of grooves or spaces through which the hot iron passes in succession, growing smaller and longer at each passage. The cylinder plies through these with a speed and force that scatters the remaining dross over the building in a brilliant shower of sparks, and at the twelfth passage comes out a bar of iron, say twenty feet long by six inches wide and one thick.

Here it ought to be remarked that the iron thus produced is puddled iron, and of inferior quality. This is to be heated and rolled, over and over, and the quality each time improved and you will please understand this to have been done with such iron as required it, and we will now follow the trail of iron to...

THE SHEARS

You may have seen a pair of scissors that would cut copper and iron wire, off, but I doubt whether you have ever seen such shears as these. A bar of railroad iron, cold, weighing seventy-three pounds to the yard, is cut off by them so easily and with as little apparent exertion as a thread by a lady's scissors. That slow, solemn crunch, of the jaw of those shears, going remorselessly through bars of iron of all sizes, lands without trembling a hair's breadth, or even going slower or hesitating, will haunt me in sleep for years.

I should have remarked that the bars of iron delivered to the Shears for the purpose of railroad iron, are of three sizes, which I give by guess work, say six inches by one, three by one, and two by one. The bars, six by two, are of the finest quality of iron...The Shears cut them into equal lengths, of about five feet, and these lengths are passed along to the hands who pile them.

This process consists simply in laying the bars in order for the rolling. A package is made, consisting of one six inch piece at the bottom, and one at the top, while the three and two inch pieces are laid alternately making a package six by six.

The packages thus made up are placed in the heating furnaces, where they are raised to a somewhat higher than welding heat, and when thus heated, are brought out to the rollers. This set of rollers of course is entirely different in its moulds from the former mentioned, and the package through four or five of them, reduced a little in size retaining its square shape, and becoming thoroughly welded. On the fifth or sixth passage it is materially reduced in size, and now assumes the railroad iron shape, which it retains, growing smaller and smaller, till on the twelfth passage, and a minute and a half from the time it commenced the first, it flies out of the rollers a perfect bar of railroad iron, and is received on a long low track, which carries it swiftly to a platform, where it is rolled over once or twice, and thus made comparatively straight, when the platform suddenly moves a few inches, presenting the bar to two circular saws which are flying with inconsiderable velocity on a shaft or axle six yards long. These saws take off the ragged ends of the bar with the velocity and ease of ordinary sawing wood. Incredible as this may seem, it is true that in thirty seconds precisely, these saws take off the two ends of a bar of railroad iron weighing seventy-three pounds to the yard, (the weight which they are now making here.) This is done entirely by the eye, and the wonderful part of it is the gigantic strength of the men engaged in it. Two of them lift a bar (weight 438 lbs.) and lay it on the supports. One now takes the end of the bar, and another mounting the platform, lifts the sledge, which weighs eighty-three pounds!

The one holding the bar draws it toward him on the rollers, or pushes it from him, occasionally giving it a slight bounce, throwing it up, turning it over, now on this side, now on that, while as he gives the word, that tremendous sledge falls on this, that, and the other place, with a light or heavy blow as he orders, until the bar is straight, and the two lift it away (to receive its final cutting and chipping at the ends,) as easily as if it weighed forty instead of four hundred pounds. I cannot tell you the gracefulness and ease with which those men swing the enormous hammers, or guide the heavy bars. They are erect, stout, noble looking fellows, all of them, and manifestly proud of their great strength. I saw them testing a bar of iron, and the rapidity and strength of their blows would have done the heart of an old Roman good, had he witnessed them. It lacks now but the finishing of the ends with chisel and file, the work of a few moments, and it is lifted to the car on which it rests, until delivered at any point between Dunkirk and New York.

But the brilliancy of all this work no one can describe. Everywhere you see flashing bars of iron, lumps of glowing metal, packages going to the rollers, bars coming from the rollers, furnaces and fires all around you, making a scene in the night time (which is the best time for a visit, the mill working steadily all night,) which is terrible, while it is grand. The one crowning beauty is the effect of the circular saws cutting off the ends of the hot bars. The finest fire wheel you have ever seen is not a tithe so brilliant and beautiful as the two circles of fire which the saws throw off.

The mill is now capable of turning off something over forty tons of railroad iron per day, and new sets of rollers are ready to be put up, which will increase the ability of the Company and the amount of iron handled, lifted, and moved in a day's work in the mill.

You have a sketch of what is going on in the heart of Pennsylvania, and one of the great sources of the future magnificence of our country. I cannot doubt that this valley is to be a great city. It will swarm like a bee hive before ten more years are passed over it. Next summer the Wyoming House will be filled with curious visitors. Already hundreds return this way from summer tours, and rejoice in the delicious air of the Wyoming Valley.

W.

Further Reading

The Scranton Iron Furnaces are discussed in varying degrees of detail in nearly all of the published works dealing with the history of Lackawanna County, Luzerne County, the city of Scranton or the Lackawanna Valley. Some of the better known of these are John Beck, *Never Before in History: The Story of Scranton* (Northridge, Calif., 1986); Burton Folsom, *Urban Capitalists* (Baltimore, 1981); Daniel Hodas, *The Business Career of Moses Taylor* (New York, 1976); H. Hollister, *History of the Lackawanna Valley* (New York, 1869); Thomas Murphy, *Jubilee History of Lackawanna County* (Topeka, 1928), J. A. Clark, *The Wyoming Valley* (Scranton, 1875), and David Craft, *History of Scranton, Penn.* (Dayton, 1891). In addition, W. David Lewis has written a thorough-going article about the company's early business dealings titled, "The Early History of the Lackawanna Iron and Coal Company: A Study in Technological Adaptation," *Pennsylvania Magazine of History and Biography* (October, 1982).

General works on the manufacture of iron and steel in America during the nineteenth century include Frederick Overman, *The Manufacture of Iron* (Philadelphia, 1850), John B. Pearse, *A Concise History of the Manufacture of Iron in the American Colonies up to the Revolution and in Pennsylvania until the Present Time* (Philadelphia, 1876), and James Swank, *History of the Manufacture of Iron in all Ages and Particularly in the United States From Colonial Times to 1891* (Philadelphia, 1892). A more recent work by Lance Metz and Craig Bartholomew titled, *Anthracite Iron Industry in the Lehigh Valley* (Easton, 1989), provides a great deal of information about the development and technology of making anthracite iron.